International Environmental Labelling

Vol.11 of 11

For All People who wish to take care of Climate Change
Tourism Industries
(Airline Industry, Travel Agent, Car Rental, Water Transport, Coach Services, Railway, Spacecraft, Hotels, Shared Accommodation, Camping, Bed & Breakfast, Cruises, Tour Operators)

Jahangir Asadi
Vancouver, BC CANADA

Suggest an ecolabel

If you think that we missed a label and/or you are an ecolabelling body, please consider to submit for the next editions of our 11 Volumes International Eco-labelling Book series. Please send your details, and we'll review your suggestions. Our goal is to be as comprehensive as possible, so thank you for your help!
info@TopTenAward.Net

Copyright © 2022 by Top Ten Award International Network.

All rights reserved. No part of this publication may be reproduced, distributed or transmitted in any form or by any means, including photocopying, recording, or other electronic or mechanical methods, without the prior written permission of the publisher, except in the case of brief quotations embodied in critical reviews and certain other noncommercial uses permitted by copyright law. For permission requests, write to the publisher, addressed "Attention: Permissions Coordinator," at the address below.

Published by: Top Ten Award International Network
Vancouver, BC **CANADA**
Email: Info@TopTenAward.net
www.TopTenAward.net

Ordering Information:
Quantity sales. Special discounts are available on quantity purchases by universities, schools, corporations, associations, and others. For details, contact the "Sales Department" at the above mentioned email address.

International Environmental Labelling Vol.11/J.Asadi—2nd ed.
ISBN 978-1-7775268-2-5

Contents

About TTAIN 10
Introduction 13
General principles of environmental labelling 20
Types of environmental labelling 24
Types I environmental labelling 28
Types II environmental labelling 46
Types III environmental labelling 52
All about 'Eco-Tourism' 54
Principles of Eco-Tourism 63
TTAIN Pioneers 80
Bibliography 85
Search by logos 92
Paper Made out of Algae 96
Environmental friendly photos 98

I dedicate this book to my friend, Habib

Acknowledgements:

I wish to thank my great friend, Mr. Habib Jamshidian who is the manager at H&B Canada Immigration Services for all of his support and sponsorship.

It should be noted that all the required permissions for using the logos and trade marks has been obtained to be published in this volume.

About TTAIN

Top Ten Award International Network

Top Ten Award international Network (TTAIN) was established in 2012 to recognize outstanding individuals, groups, companies, organizations representing the best in the public works profession.

TTAIN publishing books related to international Eco-labeling plans to increase public knowledge in purchasing based on the environmental impacts of products.

Top Ten Award International Network provides A to Z book publishing services and distribution to over 39,000 booksellers worldwide, including Apple, Amazon, Barnes & Noble, Indigo, Google Play Books, and many more.

Our services including: editing, design, distribution, marketing
TTAIN Book publishing are in the following categories:

Student
Standard
Business
Professional
Honorary

We focus on quality, environmental & food safety management systems , as well as environmnetal sustain for future kids. TTAIN also provide complete consulting services for QMS, EMS, FSMS, HACCP and Ecolabeling based on international standards.

ISO 14024 establishes the principles and procedures for developing Type I environmental labelling programmes, including the selection of product categories, product environmental criteria and product function characteristics, and for assessing and demonstrating compliance. ISO 14024 also establishes the certification procedures for awarding the label.

TTAIN has enough experiences to help create new ecolabeling programmes in different countries all over the world.
For more detail visit our website : http://toptenaward.net
and/or send your enquiery to the following email:
info@toptenaward.net

CHAPTER 1

Introduction

This book is dedicated to the subject of environmental labels. The basis for the classification of its parts goes back to the types of environmental labelling according to the classifications provided by the International Organization for Standardization. In each section, while presenting the relevant definitions, I mention the existing international standards and present examples related to each type of labelling. Environmental labelling is an important and significant topic, and its richness is added to every day, which has attracted the attention of many experts and researchers around the world. The idea of compiling this book, came to my mind when I observed that national environmental labelling models have been developed in most countries of the world, but in many other countries, the initial steps have not been taken yet. Therefore, I decided to create the first spark for the development of environmental labelling patterns in other countries by collecting appropriate materials and inserting samples of labelling patterns of different countries of the world. It should be noted that the description of each environmental label in this book does not indicate their approval or denial; they are included only to increase the awareness of all enthusiasts and consumers of the meanings and concepts derived from such labels. We hereby ask all interested parties around the world who wish to start an environmental labelling program in their country to benefit from our intellectual assistance and support in the form of consulting contracts. Increasing human awareness of the urgent need to protect the environment has led to changes in all levels of activities, including the production of marketing products, consumption, use, and sale of goods and services at the national and international levels. Stakeholders involved in environmental protection include consumers, producers, traders, scientific and technological institutes, national authorities, local and international organizations, environmental gatherings, and human society in general. Decisions by consumers and sellers of products are made not only on the basis of key points such as quality, price, and availability of

products but also on the environmental consequences of products, including the consequences that a product can have before, after and during production. The most important environmental consequences include water, soil, and air pollution along with waste generation, especially hazardous waste. Further consequences include noise, odor, dust, vibration, and heat dissipation as well as energy consumption using water, land, fuel, wood, and other natural resources. There are further effects on certain parts of the ecosystem and the environment. In addition, the environmental consequences not only include the natural use of the products but also abnormal and even emergency or accidental uses. The basis of studies and studies in this field is done through product life cycle evaluation, which generally involves the study and evaluation of environmental aspects and consequences of a category (product, service, etc.) because of the preparation of raw materials for production until they are used or discarded. Sometimes the phrase "review from cradle to grave" is used for such an evaluation. In addition to the above, the environmental consequences that may occur at any stage of the product life cycle, including the preliminary stages and its preparation, production, distribution, operation, and sale, should also be considered when evaluating it. This type of evaluation refers to product life cycle analysis from an environmental point of view,"

which is a useful tool for measuring the degree of environmental health of a product, comparing different products, improving product quality, and confirming the environmental health claims of the product. The environmental health analysis tool for products and services facilitates their placement in domestic or foreign markets, considering that the awareness of consumers and retailers about the environmental consequences of the product has increased, as has the accurate and explicit measurement by the people in charge at all levels. Local, national, and international in the field of environmental protection. Products that can claim to be environmentally complete in all stages of their life cycle and meet the mandatory and optional environmental needs are considered successful products. Environmental messages refer to the policies, goals, and skills of product manufacturing companies as part of the environmental management systems in which they are applied, and consumers and retailers are increasingly paying attention to this issue when making purchasing decisions. In addition, companies have been encouraged and even forced to adapt their environmental management systems to agencies and retailers and to local, national, international, and other environmental issues.

The environmental health message of a product can be conveyed to the consumer in various ways, including implicitly or explicitly. For example, the implicit or implicit message conveyed directly by the product to the customer is that the product is suitable for the intended use and purpose, and, without material waste in size, weight, and dimensions, is perfectly proportioned and without additional packaging. Sometimes it is necessary to convey these messages and claims about the correctness of the product quite clearly through magazines or other media as well as through certificates that are accurate, simple, and convincing to the consumer in the form of a label. These messages must be accurate and fact-based; otherwise they will nullify the product and create contradictory effects. Confirmation of these claims by a third-party organization will increase its credibility. It should also be noted that the multiplicity of these messages, depending on the type of products or companies producing them, confuses consumers in the market and also creates artificial boundaries or causes a differentiated distinction against certain products or companies. Various models, principles, and methods have been provided by local, regional, national, and international organizations to demonstrate product life cycle analysis and other guidelines on environmental management systems and their labels. At the national level, significant advances have been made in the design of environmental labels in various countries, including developing countries and the Scandinavian countries. For example, the first project was designated in Germany as a Blue Angel in 1977, later on Canada in 1988, the Scandinavian countries and Japan in 1989, the United States and New Zealand in 1990, India, Austria, and Australia in 1991, And in 1992, Singapore, the Republic of Korea, and the Netherlands developed their national environmental labelling. Environmental labels are an environmental management tool that is the subject of a series of ISO 14000 standards. These environmental labels provide information about a product or commodity in terms of its broad environmental characteristics, whether it is about a specific environmental issue or about other characteristics and topics.Interested and pro-environmental buyers can use this information when choosing products or goods. Product makers with these environmental labels hope to influence people's purchasing decisions. If these environmental labels have this effect, the share of the product in question can increase, and other suppliers may create healthy environmental competition by improving the environmental aspects of their products and commodities. The overall goal of environmental labels is to convey acceptable and accurate information that is in no way misleading regarding the environmental aspects of products and commodities, and they encourage the consumer to buy and produce products that reduce stress on the environment. Environmental labelling must follow the general principles that the International Organization for Standardization has published in a collection entitled the ISO 14020 standard, which refers to these general principles here. It should be noted that other documents and laws in this field are considered if they are in accordance with the principles set out in ISO 14020.

What is eco traveling?

Eco tourism is responsible travel to natural areas that conserves the environment and sustains the well-being of local people. According to the International Ecotourism Society, the definition of ecotourism is tourism that follows these principles: minimize impact. build environmental and cultural awareness and respect.

CHAPTER 2

General Principles on Environmental Labelling

1 The First Principle: Evironmental notices and labels must be accurate, verifiable, relevant, and in no way misleading and/or deceptive.

2 The Second Principle: Procedures and requirements for environmental labels will not be ready for selection unless they are implemented by affecting or eliminating unnecessary barriers to international trade.

3 The Third Principle: Environmental notices and labels will be based on scientific analysis that is sufficiently broad and comprehensive, and to support this claim, the product must be reliable and reproducible.

4 The Fourth Principle: The process, methodology, and any criteria required to support the announcements on environmental labels will be available upon request all interested groups.

5 The Fifth Principle: Development and improvement of environmental notices and labels should be considered in all aspects related to the service life of the product.

6 The Sixth Principle: Announcements on environmental labels will not prevent initiative and innovation but will be important in maintaining environmental implementation.

7 The Seventh Principle: Any enforcement request or information requirement related to environmental notices and labels should be limited to the necessary information to establish compliance with an acceptable standard and based on the notification standards and environmental labels.

8 The Eighth Principle: The process of improving the announcement and environmental labels should be done by an open solution with interested groups. Reasonable impressions must be made to reach a consensus through this process.

9 The Ninth Principle: Information on the environmental aspects of the product and goods related to an advertisement and environmental label will be prepared for buyers and interested buyers from a group consisting of an advertisement and an environmental label.

What is the other name for ecotourism?
You can discover synonyms, antonyms, idiomatic expressions, and related words for eco-tourism, like: tourism, environmental-conservation, ecotourism, quichua and wildlife-conservation.

CHAPTER 3

Types of Environmental Labelling

At present, according to the classification provided by the International Organization for Standardization, there are three types of environmental labelling patterns:

1. Type I labelling: This labelling is known as eco-labelling, and because it is difficult to translate this word into many languages, it presents another reason to adhere to a numerical classification system. In the content of Type I labelling, a set of social commitments that creates criteria according to the scientific principles on the basis of which a product is environmentally preferable is discussed. Consumers are then instructed in assessing environmental claims and must decide which packaging is more important.

2. Type II labelling: refers to the claims made on product labels in connection with business centers. This includes familiar claims such as recyclable, ozone-friendly, 60% phosphate-free, and the like. This type of labelling can be in the form of a mark or sentence on the product packaging. Some of them are valid environmental claims—and some can be completely misleading. Usually, all countries have laws against deceptive advertisements, so why has the International Organization for Standardization discussed this issue? The answer is that it is not clear whether the environmental claims have a technical basis or whether the ad is meaningless.

3 Type III labelling: is a distinct form of third-party environmental labelling pattern designed to avoid the difficulties that can result from type-one labelling. Technical committee for Environment of International organization for Standardization has undertaken a new project to standardize guidelines and Type III labelling methods. One of the main objections raised by industries to Type I labelling is the basis for its management.

What is Ecotourism?
Ecotourism is defined as "responsible travel to natural areas that conserves the environment, sustains the wellbeing of local people and involves interpretation and education"
(International Ecotourism Society, 2015)

CHAPTER 4

Type I Environmental Labelling

Type I labelling: This labelling is known as eco-labelling, and because it is difficult to translate this word into many languages, it presents another reason to adhere to a numerical classification system. In the content of Type I labelling, a set of social commitments that creates criteria according to the scientific principles on the basis of which a product is environmentally preferable is discussed. Consumers are then instructed in assessing environmental claims and must decide which packaging is more important.

Type I adhesive has the following specifications:
A. Has an optional third-party template.
B. When the product meets a certain standard, the labelling of this product is included.
C. The purpose of this program is to identify and promote products that play a pioneering role in terms of environment, which means its criteria are at a higher level than the average environmental performance.
D. Acceptance/rejection criteria are determined for each group of products and are publicly available.
E. The criteria are adjusted after considering the environmental consequences of the product life cycle.

Examples of Type I Labelling:
In this section, and considering the importance of this type of labelling, I provide a description of some examples of Type I labelling related to some countries along with a list of products on which this mark is placed.

Germany

FSC® is a global not-for-profit organization that sets the standards for responsibly managed forests, both environmentally and socially. When timber leaves an FSC certified forest they ensure companies along the supply chain meet our best practice standards also, so that when a product bears the FSC logo, you can be sure it's been made from responsible sources. In this way, FSC certification helps forests remain thriving environments for generations to come, by helping you make ethical and responsible choices at your local supermarket, bookstore, furniture retailer, and beyond. www.fsc.org

FSC® International
Adenauerallee 134
53113 Bonn
E-mail: info@fsc.org
Phone: +49 (0) 228 367 66

FSC Canada
50 rue Sainte-Catherine Ouest,
bureau 380B, Montreal, QC H2X 3V4
Email: info@ca.fsc.org
Telephone: 514-394-1137

Ukraine

The ecolabelling program in Ukraine was founded on the initiative of the All-Ukrainian NGO "Living Planet" in 2003. The Green Crane is the first and the only one Type 1 Ecolabel in Ukraine that recognized officially.

The main objective of company's activity is to evaluate the products for compliance with environmental criteria according to ISO 14024 scheme in order to ensure the reliability of data on the environmental benefits of products within a specific category based on the results of the life cycle assessment. Over the 16 years of the program's existence, the Green Crane ecolabel has become a recognizable reliable reference point for consumers and government organizations (in "green procurement" process), as well as effective marketing tool for business.

Program Statistical Information. Today, the program operates with 57 certification standarts in various industries - construction, food, chemical, textile and other. More than 500 certificates have been issued throughout the program's history. Currently, 68 certificates for more than 1,000 products are valid.

Contact:
NGO «Living Planet»
Email: os@ecolabel.org.ua,
              info@ecolabel.org.ua
Tel: +380 44 332 84 08
Adress: Magnitogorsky Lane1-B, Kyiv, Ukraine - 02094
Web: https://www.ecolabel.org.ua/en

Green Key

Global

The Green Key award is the leading standard for excellence in the field of environmental responsibility and sustainable operation within the tourism industry. This prestigious eco-label represents a commitment by businesses that their premises adhere to the strict criteria set by the Green Key programme under the auspices of the Foundation for Environmental Education (FEE). Guests who opt to stay in Green Key awarded establishments can know that they are they are helping to make a difference on an environmental level. The high environmental standards expected of these establishments are maintained through rigorous documentation and frequent audits. Green Key is eligible for hotels, hostels, small accommodations, campsites, holiday parks, conference centres, restaurants and attractions.

Contact:
Foundation for Environmental Education
Scandiagade 13,
2450 Copenhagen SV, DENMARK
Website: www.greenkey.global
E: info@fee.global

Hong Kong

The Green Council is a non-profit, tax-exempt charitable environmental stewardship organisation and certification body (Reg. No.: HKCAS-027) of Hong Kong established in 2000. A group of individuals from different sectors of industry and academics shared the vision to help build Hong Kong into a world-class green city for the future. They formed the Green Council with the aim of encouraging the commercial and industrial sectors to include environmental protection in their management and production processes. The Green Council is a non-profit, tax-exempt charitable environmental stewardship organisation and certification body (Reg. No.: HKCAS-027) of Hong Kong established in 2000. A group of individuals from different sectors of industry and academics shared the vision to help build Hong Kong into a world-class green city for the future. They formed the Green Council with the aim of encouraging the commercial and industrial sectors to include environmental protection in their management and production processes. The Green Council is a non-profit, tax-exempt charitable environmental stewardship organisation and certification body (Reg. No.: HKCAS-027) of Hong Kong established in 2000. A group of individuals from different sectors of industry and academics shared the vision to help build Hong Kong into a world-class green city for the future. They formed the Green Council with the aim of encouraging the commercial and industrial sectors to include environmental protection in their management and production processes.

Contact:
Website: https://www.greencouncil.org/hkgls
Email: info@greencouncil.org
Telephone: (852) 2810 1122

Global

Recognized globally and trusted by millions around the globe, the iconic Blue Flag is one of the world's most renowned voluntary awards for beaches, marinas, and sustainable boating tourism operators. Operated under the auspices of the Foundation for Environmental Education (FEE), sites qualifying for the prestigious Blue Flag award are required to meet and maintain a series of stringent environmental, educational, safety, and accessibility. Central to the ideals of the Blue Flag programme is the aim of connecting the public with their surroundings and encouraging them to learn more about their environment.

Contact:
Foundation for Environmental Education
Scandiagade 13,
2450 Copenhagen SV, DENMARK
Website: www.blueflag.global
E: info@fee.global

EUROPE

Established in 1992 and recognized across Europe and worldwide, the EU Ecolabel is a label of environmental excellence that is awarded to products and services meeting high environmental standards throughout their life-cycle: from raw material extraction, to production, distribution and disposal. The EU Ecolabel promotes the circular economy by encouraging producers to generate less waste and CO_2 during the manufacturing process. The EU Ecolabel criteria also encourages companies to develop products that are durable, easy to repair and recycle.

The EU Ecolabel criteria provide exigent guidelines for companies looking to lower their environmental impact and guarantee the efficiency of their environmental actions through third party controls. Furthermore, many companies turn to the EU Ecolabel criteria for guidance on eco-friendly best practices when developing their product lines. The EU Ecolabel helps you identify products and services that have a reduced environmental impact throughout their life cycle, from the extraction of raw material through to production, use and disposal. Recognised throughout Europe, EU Ecolabel is a voluntary label promoting environmental excellence which can be trusted.

Spain , Germany, Italy, Sweden, Greece, Portugal, Poland, Belgium, Netherlands, Estonia, Finland, Austria, Lithuania, Czech Republic, Norway, Cyprus, Ireland, Slovenia, Hungary, Romania, Croatia, Bulgaria, Malta, Slovak Republic, Latvia, Luxembourg, Iceland

Contact and more information via: http://ec.europe.eu

Global

Recognised by UNESCO and UN Environment as a world-leader within the field of Education for Sustainable Development, the Eco-Schools programme is one of five programmes run by the Foundation for Environmental Education (FEE). Based on a 7-Step Methodology, the Eco-Schools programme encourages young people to engage in their environment through project-based, experiential learning, focused on positive sustainable actions. With over 59,000 schools in 72 countries, Eco-Schools is the largest global sustainable schools programme and has helped shaped millions of young people into sustainably-minded, environmentally conscious individuals for more than 25 years.

Contact:
Foundation for Environmental Education
Scandiagade 13,
2450 Copenhagen SV, DENMARK
Website: www.ecoschools.global
E: info@fee.global

Republic of Korea

The Korea Eco-labelling is a certification system enforced by the Ministry of Environment and KEITI(Korea Environmental Industry & Technology Institute). Since its foundation in April 1992, the system has certified a wide range of eco-friendly products, which were selected as excellent not only in terms of their environmental-friendliness, but also for their quality and performance during their life cycle. Korea Eco-labelling is voluntary certification scheme to attach logo to products with superior environmental quality throughout their lifecycle to other products of the same use, and thus to provide product information to consumers. For 30 years, the scheme has launched plenty of eco-labelling product standards covering personal and household goods, construction materials, office equipment furniture, etc. It products categories which cover all aspects of products, such as reduction of use of harmful substances, energy saving, resource saving, etc. As of April 30th 2021, 169 criterias(=standards), and certifications for 18,250 products(4,549 companies) have maintained.

Contact:
Korea Environmental Industry & Technology Institute(KEITI)
Office of Korea Eco-Label Innovation
Address: 215, Jinheung-ro, Eunpyeong-gu, Seoul, Repulic of Korea
T: +82 2 2284 1518
F: +82 2 2284 1526
E: accolly@keiti.re.kr
W: www.keiti.re.kr

Peru

BIO LATINA, the consolidated byproduct of four Latin American national certification entities. Since 1998, we have provided certification services in Latin America for national and international markets. We seek to help create a more sustainable and resilient world. With these goals in mind, we have expanded our service portfolio beyond organic to social and environmental certifications.

Visit us: https://biolatina.com

From our regional offices we serve Latin American.

Our headquaters:
Av. Javier Prado Oeste 2501, Bloom Tower Of. 802, Magdalena del Mar, Lima 17, Perú

ORGANIC CERTIFICATION

Lithuania

EKOAGROS is the only institution in Lithuania for more than 20 years carrying out certification and control activities of organic production and products of national quality, also providing services of certification activities in accordance with the foreign national and private standards in foreign countries. From year 2017 EKOAGROS is accredited as certifying agent to conduct certification activities on crops, wild crops, livestock and handling operations in accordance with USDA NOP.

Contact information:
EKOAGROS
Address K. Donelaicio str. 33, LT-44240 Kaunas, Lithuania
Tel. No. +370 37 20 31 81
Website: www.ekoagros.lt

Global

The Learning about Forests (LEAF) programme advocates for outdoor learning and hands-on experiences that offer children and youth a deeper and more involved understanding of the natural world. Under the auspices of the Foundation for Environmental Education (FEE), the LEAF programme is being implemented in over 25 countries. While the focus of the LEAF programme is on tree-based ecosystems, the skills and knowledge acquired can be applied to any natural environment!

Contact:
Foundation for Environmental Education
Scandiagade 13,
2450 Copenhagen SV, DENMARK
Website: www.leaf.global
E: info@fee.global

Thailand

The Thai Green Label Scheme was initiated by the Thailand Business Council for Sustainable Development (TBCSD) in October 1993. It was formally launched in August 1994 by The Thailand Environment Institute (TEI) and Thai Industrial Standards Institute (TISI). The Green Label is an environmental certification logo awarded to specific products which have less detrimental impact on the environment in comparison with other products serving the same function. The Thai Green Label Scheme applies to all products and services, but not foods, beverage, and pharmaceuticals. Products or services which meet the Thai Green Label criteria may carry the Thai Green Label. Participation in the scheme is voluntary.

Thailand Environment Institute (TEI)
16/151 Muang Thong Thani, Bond Street,
Bangpood, Pakkred, Nonthaburi 11120 THAILAND
Tel. +66 2 503 3333 ext. 303, 315, 116
Fax. +66 2 504 4826-8
Website: http://www.tei.or.th/greenlabel/
Email: lunchakorn@tei.or.th

Netherland

For more than 25 years, the independent Dutch foundation SMK works from professional knowledge with companies to improve the sustainability of products and business management. SMK cooperates with an extensive stakeholder network of governments, producers, branch and non-governmental organisations, retailers, consultancies, researchers. The SMK Boards of Experts establish objective criteria for more sustainable products and services. SMK's transparent work processes, third party audits and certifications are conducted according to international certification standards, mostly under supervision of the Dutch Accreditation Council. Besides, SMK is Competent Body of the EU Ecolabel. SMK keeps an extensive database of sustainability criteria.

Contact:
Bezuidenhoutseweg 105 - 2594 AC Den Haag
Telefoon: 070-3586300
Mobiel: 06-82311031
(niet op woensdag)
www.smk.nl

Young Reporters for the environment

Global

The Young Reporters for the Environment (YRE) programme is an award-winning programme coordinated by Foundation for Environmental Education (FEE). Since 1994, the YRE programme has trained over 450,000 youth as Young Reporters for the Environment. Young Reporters produce high-quality journalistic work - articles, videos and photographs - that raise awareness of environmental issues in their local communities and advocate for sustainable solutions. The programme empowers young people to take an educated stand on environmental issues they feel strongly about and gives them confidence to become leaders on these issues.

Contact:
Foundation for Environmental Education
Scandiagade 13,
2450 Copenhagen SV, DENMARK
Website: www.yre.global
E: info@fee.global

USA

The Carbonfree® Product Certification is a meaningful, transparent way for you to provide environmentally-responsible, carbon neutral products to your customers. By determining a product's carbon footprint, reducing it where possible and offsetting remaining emissions through our third-party validated carbon reduction projects, companies can:
- Differentiate their brand and product
- Increase sales and market share
- Improve customer loyalty
- Strengthen corporate social responsibility & environmental goals

The Carbonfree® Product Certification Program is proud to be part of Amazon's Climate Pledge Friendly Program!
Carbonfund.org is leading the fight against climate change, making it easy and affordable to reduce & offset climate impact and hasten the transition to a clean energy future.

Contact:

O: 240.247.0630 ext 633
C: 203.257.7808
M: 853 Main Street, East Aurora, NY, 14052

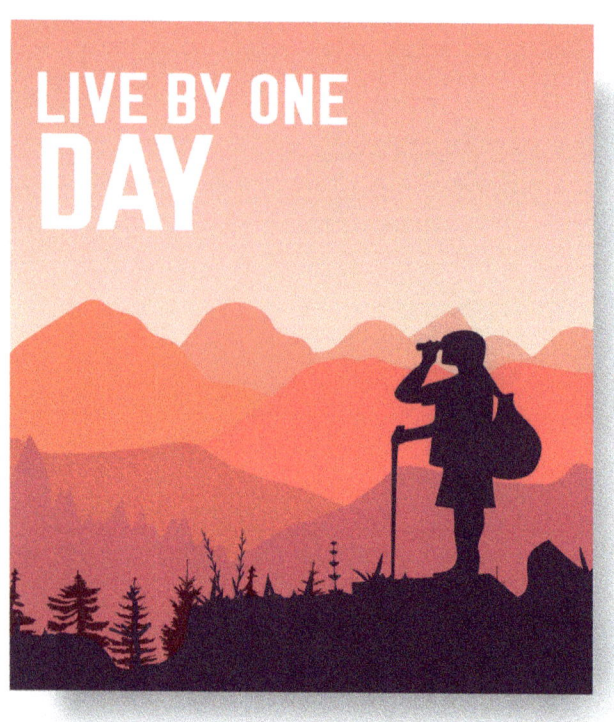

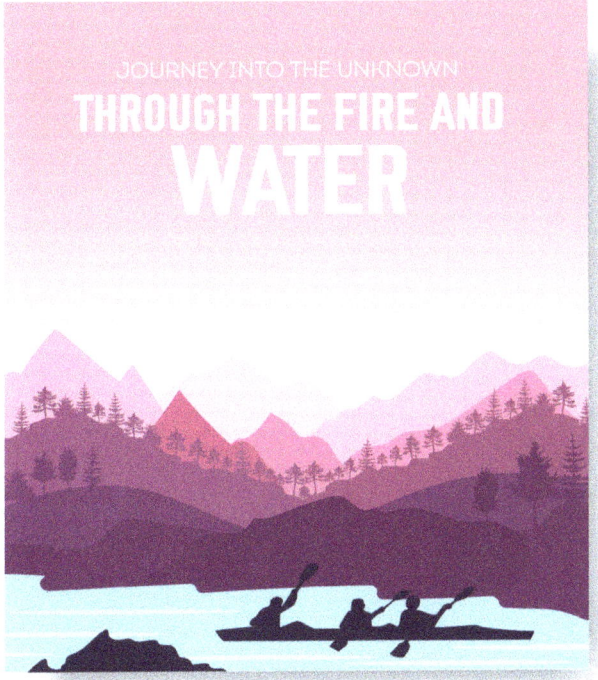

CHAPTER 5

Type II Environmental Labelling

Type II environmental labelling refers to the claims made on product labels in connection with business centers. This includes familiar claims such as recyclable, ozone-free, 60% phosphate-free, and the like. This type of labelling can be in the form of a mark or sentence on the product packaging. Some of them are valid environmental claims—and some can be completely misleading.

Usually, all countries have laws against deceptive advertisements, so why has the International Organization for Standardization discussed this issue? The answer is that it is not clear whether the environmental claims have a technical basis or whether the ad is meaningless.

Most countries have guidelines at the national level to help producers and consumers know what constitutes a true, scientifically valid claim.
There is a national standard on this in Canada. In Australia, the Consumer Commission has published guidance on this, and there are similar examples in other countries.

Global

Environmental Sustain for Future kids established in Vancouver, BC Canada in 2020. (ESFK) is an international ecolabel focused on taking care of environment for future of kids.

ESFK defined as 'self-declared' environmental claims made by manufacturers and businesses based on ISO 14020 series of standards, the claimant can declare the environmental objectives and targets in relation to taking care of environment for future kids. However, this declaration will be verifiable.

Environmental Sustain for Future Kids
Vancouver, BC CANADA

Email: info@esfk.org
Web: www.esfk.org

For all People who wish to take care of Climate Change

For all Schools, Libraries, Homes and/or Offices

Available in more than 39,000 booksellers worldwide, including Amazon, Barnes & Noble, Google Play Books, Walmart, ... and many more.

International Environmental Labelling
Set Box Book series (Vol.1-11)
+ Free Knowledge Test

Special Thanks to:
United Nations Environment Programme
(UNEP)
& more than 57 Countries

Order Now
online

http://TopTenAward.Net

CHAPTER 6

Type III Environmental Labelling

Type III environmental labelling is a distinct form of third-party environmental labelling pattern designed to avoid the difficulties that can result from type I labelling. Technical committee for Environment of International organization for Standardization has undertaken a new project to standardize guidelines and Type III labelling methods. One of the main objections raised by industries to Type I labelling is the basis for its management.

Due to the nature of the system, less than 50% of the various products on the market can meet the criteria and qualify for Type I Labelling. As long as the industry is the main supporter of other third-party models for quality systems, it is sometimes difficult for an industry to support a program that can only benefit 15% of its members. This type of labelling is currently practiced in some countries, such as Sweden, Canada, and the United States. Choosing the right product has never been easy, but Type III labelling will help because each product can have a label that describes its environmental performance and is certified by a third-party company. Consumers can then compare labels and choose their favorite products.

CHAPTER 7

All about 'Eco-Tourism'

The term Ecotourism emerged in the late 1980s as a direct result of theworld's acknowledgment and reaction to sustainable practices and global ecological practices. In these instances, the natural-based element of holiday activities together with the increased awareness to minimise the 'antagonistic' impacts of tourism on the environment (which is the boundless consumption of environmental resources) contributed to the demand for ecotourism holidays. This demand was also boosted by concrete evidence that consumers had shifted away from mass tourism towards experiences that were more individualistic and enriching. In addition, these experiences were claimed to be associated with a general search for the natural component during holidays

Definitions of Ecotourism

Ziffer, 1989	'Ecotourism is a form of tourism inspired primarily by the natural history of an area, including its indigenous cultures. The ecotourist visits relatively undeveloped areas in the spirit of appreciation, participation and sensitivity. The ecotourist practices a non-consumptive use of wildlife and natural resources and contributes to the visited area through labor or financial means aimed at directly benefiting the conservation of the site and the economic well-being of the local residents...'
Boo, 1991	'Ecotourism is a nature tourism that contributes to conservation, through generating funds for protected areas, creating employment opportunities for local communities, and offering environmental education.'
Forestry Tasmania, 1994	'Nature-based tourism that is focused on provision of learning opportunities while roviding local and regional benefits, while demonstrating environmental, social, cultural, and economic sustainability'
Richardson, 1993	'Ecologically sustainable tourism in natural areas that interprets local environment and cultures, furthers the tourists' understanding of them, fosters conservation and adds to the well-being of the local people.'
Australia Department of Tourism, 1994	'Nature-based tourism that involves education and interpretation of the natural environment and is managed to be ecologically sustainable. This definition recognizes that natural environment includes cultural components, and that ecologically sustainable involves an appropriate return to the local community and long-term conservation of the resource.'
Figgis, 1993	'Travel to remote or natural areas which aims to enhance understanding and appreciation of natural environment and cultural heritage, avoiding damage or deterioration of the "environment and the experience for others".'
Tickell, 1994	'Travel to enjoy the world's amazing diversity of natural life and human culture without causing damage to either.'

Definitions of Ecotourism (Cont.)

Boyd & Butler, 1993	'A responsible nature travel experience, that contributes to the conservation of the ecosystem while respecting the integrity of host communities and, where possible, ensuring that activities are complementary, or at least compatible, with existing resource-based uses present at the ecosystem.'
Boyd & Butler, 1996	'Ecotourism is a form of tourism which fosters environmental principles, with an emphasis on visiting and observing natural areas'
Goodwin, 1996	'Low impact nature tourism which contributes to the maintenance of species and habitats either directly through a contribution to conservation and/or indirectly by providing revenue to the local community sufficient for local people, and therefore protect, their wildlife heritage area as a source of income.'
Lindberg & McKercher, 1997	'Ecotourism is tourism and recreation that is both nature-based and sustainable.'

Environmental Impacts of Ecotourism

The most proclaimed positive issue is ecotourism's contribution to sustainable resource management through conservation of the natural resources on a direct or indirect basis (Commonwealth of Australia, 1993, 1995; Cater, 1993, 1994; Dearden, 1995)

Environmental impacts	
Direct benefits	Direct costs
• Provides incentive to protect environment, both formally (protected areas) and informally	• Danger that environmental carrying capacities will be unintentionally exceeded, due to:
• Provides incentive for restoration and conversion of modified habitats	• Rapid growth rates Difficulties in identifying, measuring and monitoring impacts over a long period
• Ecotourists actively assisting in habitat enhancement (donations, policing, maintenance, etc.)	• Idea that all tourism induces stress

Environmental impacts (Cont.)	
Indirect benefits	Indirect costs
• Exposure to ecotourism fosters broader commitment to environmental well-being	• Fragile areas may be exposed to less benign forms of tourism (pioneer function)
• Space protected because of ecotourism provide various environmental benefits	• May foster tendencies to put financial value on nature, depending upon attractiveness

Economic Impacts of Ecotourism

The direct and indirect benefits which are derived from biodiversity conservation, represent the fundamental goal of ecotourism, by attracting visitors to the natural settings and using the revenues to fund conservation and fuel economic development (Commonwealth of Australia, 1995: 12; Cater, 1993, 1994)

Economic impacts	
Direct benefits	Direct costs
• Revenues obtained directly from ecotourists • Creation of direct employment opportunities • Strong potential for linkages with other sectors of the local economy • Stimulation of peripheral rural economies	• Start-up expenses (acquisition of land, establishment of protected areas, superstructure, infrastructure) • Ongoing expenses maintenance of infrastructure, promotion, wages)

Economic impacts (Cont.)	
Indirect benefits	Indirect costs
• Indirect revenues from ecotourists (high multiplier effect) • Tendency of ecotourists to patronise cultural and heritage attractions as 'add-ons' • Economic benefits from sustainable use of protected areas and inherent existence	• Revenue uncertainties to in situ nature if consumption • Revenue leakages due to imports, expatriate or non-local participation, etc. • Opportunity costs • Damage to crops by wildlife

Sociocultural Impacts of Ecotourism

The sustainable component of ecotourism often attests certain direct and indirect sociocultural benefits and costs at the sites and/or at the destination level . Generally speaking, it was proposed that the assessment of the cultural impacts of ecotourism could be based on four criteria , commodification element; culture affecting social change; cultural knowledge; and cultural patrimony elements.

Sociocultural impacts	
Direct benefits	Direct costs
• Ecotourism accessible to a broad spectrum of the population	• Intrusions upon local and possibly isolated cultures
• Aesthetic/spiritual element of experiences	• Imposition of elite alien value system
• Foster environmental wareness among ecotourists and local population	• Displacement of local cultures by parks
	• Erosion of local control (foreign experts, in-migration of job seekers).

Sociocultural impacts (Cont.)	
Indirect benefits	Indirect costs
• Option and existence benefits	• Potential resentment and antagonism of locals • Tourist opposition to aspects of local culture (e.g. hunting, slash-burn agriculture).

Principles of Ecotourism

(1) travel to natural destinations.

(2) minimizes impact. This includes minimizing the impact of development and tourist activity by choosing appropriate building materials, renewable energy sources, visitor management strategies, monitoring techniques and conservation plans.

(3) builds environmental awareness. This includes educational and interpretational material for visitors, educational training for guides and educating the greater public and surrounding community.

(4) provides direct financial benefit for conservation.

(5) provides financial benefits and empowerment for local people. This includes employment of local people, using an all-inclusive stakeholder approach to planning, management and policy development and fostering of partnerships.

(6) respects local culture.

(7) supports human rights.

LIVE BY ONE
DAY

THROUGH THE FIRE AND
WATER

RIDE ON THE WINGS OF
WIND

ALONE WITH
NATURE

CHAPTER 8

Top Ten Eco Hotels and Lodges Selected by :
Top Ten Award International Network

Top Ten Award international Network (TTAIN) was established in 2012 to recognize outstanding individuals, groups, companies, organizations representing the best in the public works profession.
TTAIN publishing books related to international Eco-labeling plans to increase public knowledge in purchasing based on the environmental impacts of products. We introduce in each volume some of the organizations that are doing their best in relation to taking care of the environmnet.
What makes a hotel an eco hotel?
An eco-friendly hotel works to reduce its environmental impact by employing sustainable best practices in maintenance, services, and supply chains. Measures may include reducing energy and water consumption, aiming for zero waste generation, and using environmentally friendly products. How do you build an eco hotel? So, if you're about to embark upon building a hotel, here are some eco-conscious suggestions.
Use Eco-Friendly Building Materials. ...
LED Lights. ...
Make Guest Rooms More Green. ...
Water Points. ...
Have Wooden Key Cards. ...
Reduce Paperwork and ...

Svart, Norway

Svart is the first building to be designed and built according to the highest energy efficiency standards in the northern hemisphere and aim to meet the Paris agreement environmental standards. We estimate the hotel to save 85% of its annual energy consumption and harvest enough solar energy to cover both the hotel operations, including its boat shuttle operation and the energy needed to construct the building. The ambition is to further our goals and develop scalable technology and operational know-how to replicate and introduce on other developments to reduce energy consumption, improve building standards and upgrade practice values and profitability.

Contact:
Address: Holandsfjorden, 8178 Halsa, Norway
E: post@svart.no

Jetwing Surf, Sri Lanka

As an eco-luxury hotel on our eastern coast, Jetwing Surf features a natural commitment to sustainability. Inspired by the resourceful landscapes that surround us, our home of Sri Lankan hospitality has been designed in harmonious coexistence – allowing us to promote, preserve, and protect our thriving environment through a number of initiatives that make Jetwing Surf a symbol of responsible tourism. Inspired by the waves and seashells of our coastal home, Jetwing Surf features a unique, open architectural design across our rooms, reception and restaurant, which reduces the requirement for artificial illumination and ventilation. None of our cabanas feature air conditioners due to the sustainable combination of an iluk roof which helps minimise heat gain during the day, a steep roof pitch which prevents stagnation of warm air in living spaces, and cadjan-covered double layered walls which not only allow for the easy flowing of fresh air, but also provide you with absolute privacy. In addition, the open-air bathrooms also require no mechanical ventilation, as every cabana at Jetwing Surf has been designed to allow fresh air from cool ocean breeze to flow through naturally.

Contact:
Jetwing Surf,
P20, Kottukal Beach Rd.
Hidayapuram,
Pottuvil,
SRI LANKA

Garonga Safari Camp, South Africa

Fall in step with nature and encounter the wildlife that roam in this South African wilderness. Game drives and walking safaris with our expert guides get you up-close to the Big 5, endangered species like cheetah and wild dog, and diverse birdlife.

From intimate wildlife encounters to personal time to reflect in nature, our abundance of relaxing safari activities is designed to encourage exactly that. Try your hand at everything, or simply unwind in a hammock, surrounded only by the sights and sounds of the reserve.

Contact:

Hoedspruit 1380, South Africa

El Nido Resorts, Philippines

Palawan is an archipelago of 1,780 islands on the western part of the Philippines. It has the most concentration of islands but is the most sparsely populated region in the country. Because of its scenic landscapes and high bio-diversity, Palawan is known as "The Last Ecological Frontier of the Philippines".

The northern part of Palawan province is blessed with crystal-clear waters, pristine beaches, and a wealth of flora and fauna. It is here that El Nido and Taytay, home of the lovely El Nido Resorts, are located. Spectacular ancient limestone cliffs tower over marine sanctuaries teeming with innumerable species of tropical fish and coral, as well as five species of endangered sea turtles. Lush forests abound with more than 100 species of birds. It is a truly exotic destination.

Contact:
Address: El Nido, Palawan 5313 Philippines

Three Camel Lodge, Mongolia

In the light of the Gobi Desert, the Three Camel Lodge appears like a mirage, nestled between the Bayanzag cliffs and the Altai Mountains. A cluster of comfortable, eco-conscious gers: round, felt tents, decorated warmly in the Mongolian tradition, with local materials and hand-painted interiors.

Yes, it looks sensational, but most importantly it's socially responsible – the solar-powered ger camp has banned plastic use, protects wildlife and uses profits to support struggling nomads in winter.

Contact:
web: https://www.threecamellodge.com/
Email: info@ threecamellodge.com

Pacuare Lodge, Costa Rica

Pacuare Lodge is a luxury, eco-hotel with only 20 suites and situated on the banks of one of the world's most scenic, white-water wonders, the Pacuare River of Costa Rica. This unparalleled location, with its vibrant nature, offers a scenic backdrop to adventure, romance and wellness.

Simple, and yet sophisticated, the indigenous Cabécar-inspired architecture, pristine rainforest, farm-to-table exquisite cuisine, and luxurious, well-appointed accommodations come together to offer an experience like no other.

The Pacuare Lodge was built with minimal impact on the river and rainforest No trees were cut to build the bungalows and facilities. Instead, lumber was purchased from a sustainable reforestation project operated by small farmers.

Contact:
https://www.pacuarelodge.com/

Bambu Indah, Indonesia

Deep immersion in our natural environment with attention to detail of a boutique hotel, A fusion of two worlds lies at Bambu Indah. Serenity and peace of mind is on full display, accompanied by warm local hospitality and luxury service to ensure the most memorable experience possible. We've combined all the elements we love about hospitality and travel to create an adventure for your senses that we feel is truly unparalleled. Peak over the volcanic rocks and catch your reflection in our natural, green swimming pond. Complete with a rope swing so you can take the ultimate plunge into nature's pool.

Contact:
Jl. Baung, Sayan, Kecamatan Ubud,
Kabupaten Gianyar,
Bali 80571, Indonesia

Karinji Eco Retreat, Western Australia

Karijini Eco Retreat is one of Australia's leading eco-tourism attractions. Situated 1,500km north of Perth in WA's second largest national park, the retreat offers safari-style eco tents, cabins and campsites nestled amongst native bushland at the edge of Joffre Gorge, along with an outback restaurant and bar, and access to the park's walks and guided adventure tours. Stay under the stars in our campground, or in comfortable glamping accommodation at the first and only facility to offer a glamping experience in the Karijini National Park. Explore a wonderland of ancient natural landscapes formed more than two billion years ago – deep gorges, red cliffs, towering waterfalls and emerald green waterholes, while experiencing a taste of genuine mateship and friendly hospitality. With its hulking gorges, thundering waterfalls and rolling hills strewn with wildflowers and wallabies, Karijini National Park is one epic spot for hiking, climbing and canyoning. What's more, this eco retreat (based in the park) treads lightly on the landscape, thanks to its solar power, lack of air conditioning and non-permanent (yet beautifully furnished) tented suites.

Contact:
Address: 1/589 Stirling Highway, Cottesloe Western Australia 6011
http://karijiniecoretreat.com.au

Rewa Lodge, Guyana

Rewa village is located in the center of Guyana, a small country in the north-eastern part of South America. Guyana is internationally recognized for its pristine natural environment and tremendous biodiversity.

Rewa and our eco-lodge are found at the confluence of the Rupununi and Rewa Rivers. The Rewa River is uninhabited and provides a unique opportunity to explore untouched habitat. Rewa Eco-lodge was built by and is run by the people of Rewa Village. This enterprise provides our people with employment opportunities while adding value to the incredible resource that is our forest's biodiversity.

Contact:
https://www.rewaecolodge.com/

Mashpi Lodge, Ecuador

Ecuador's extraordinary Mashpi Lodge is an eco-wonder set deep in the amazing Chocó-Andean Cloud Forest. At night, you can sleep well, knowing that not a single tree was removed during its construction. During all hours, you'll be treated to uninterrupted views of rainforest from your room, and the hundreds of species of birds, butterflies, and jungle frogs that call this jungle home. Located at a former lumber mill, Mashpi Lodge was built using the latest techniques in sustainable construction in order to prevent damage to the forest, with much of the structure having been preassembled in Quito and designed around the topography of the site to prevent further damage to the trees. It is designed to blend in perfectly with its natural environment, respecting the natural space it occupies. Stunningly contemporary and featuring modernist design and décor that mixes warm earth tones, steel, stone, and glass in striking perspectives, Mashpi Lodge makes for a true cocoon of luxury in the middle of the forest.

Contact:
http://mashpilodge.com
Email: info@mashpilodge.com

About UNWTO

The World Tourism Organization (UNWTO) is the United Nations agency responsible for the promotion of responsible, sustainable and universally accessible tourism.

CHAPTER 9

Top Ten Award International Network Environmental Pioneers

Top Ten Award international Network (TTAIN) was established in 2012 to recognize outstanding individuals, groups, companies, organizations representing the best in the public works profession. TTAIN publishing books related to international Eco-labeling plans to increase public knowledge in purchasing based on the environmental impacts of products. We introduce in each volume some of the organizations that are doing their best in relation to taking care of the environmnet.

Global

The Foundation for Environmental Education (FEE) is the world's largest environmental education organisation, with 100 member organisations in 82 countries. Our educational programmes, Eco-Schools, Learning About Forests and Young Reporters for the Environment, empower young people to create an environmentally conscious world through a solutions-based approach. Our Green Key and Blue Flag programmes are globally recognized for promoting sustainable business practices and the protection of natural resources. With 40 years of impactful experience in ESD, FEE's Strategic Plan, GAIA 20:30, prioritises climate action across all five programmes to address the urgent threats of climate change, biodiversity loss and environmental pollution.

Contact:
Foundation for Environmental Education
Scandiagade 13,
2450 Copenhagen SV, DENMARK
Website: www.fee.global
E: info@fee.global

UNEP

The United Nations Environment Programme (UNEP) is the leading global environmental authority that sets the global environmental agenda, promotes the coherent implementation of the environmental dimension of sustainable development within the United Nations system, and serves as an authoritative advocate for the global environment.

Our mission is to provide leadership and encourage partnership in caring for the environment by inspiring, informing, and enabling nations and peoples to improve their quality of life without compromising that of future generations.

Headquartered in Nairobi, Kenya, we work through our divisions as well as our regional, liaison and out-posted offices and a growing network of collaborating centres of excellence. We also host several environmental conventions, secretariats and inter-agency coordinating bodies. UN Environment is led by our Executive Director.

We categorize our work into seven broad thematic areas: climate change, disasters and conflicts, ecosystem management, environmental governance, chemicals and waste, resource efficiency, and environment under review. In all of our work, we maintain our overarching commitment to sustainability.

Website: www.unep.org

Bibliography

Bibliography:
Achama, F. (1995) Defining ecotourism. In L. Haysith and J. Harvey (eds) Nature Conservation and Ecotourism in Central America (pp. 23–32). Florida: Wildlife Conservation Society.
Agardy, M.T. (1993)Accommodating ecotourism in multiple use planning of coastal and marine protected areas. Ocean & Coastal Management 20 (3), 219–239.
Australia Department of Tourism (1994)National Ecotourism Strategy. Canberra: Australia Government Publishing Service.
Ayala, H. (1995) From quality product to eco-product:Will Fiji set a precedent? Tourism Management 16 (1), 39–47.
Barnes, J.L. (1996)Economic characteristics of the demand for wildlife-viewing tourism in Botswana. Development Southern Africa 13 (3), 377–397.
Blamey, R.K. (1995a) The Nature of Ecotourism. Canberra: Bureau of Tourism Research.
Blamey, R.K. (1995b) The elusive market profile: Operationalising ecotourism. Paper presented at the Geography of Sustainable Tourism Conference, University of Canberra, ACT, Australia, September.
Blamey, R.K. (1997) Ecotourism: The search for an operational definition. Journal of Sustainable Tourism 5 (2), 109–130.
Boo, E. (1990) Ecotourism: The Potential and Pitfalls (Vols 1& 2). Washington, DC: World Wide Fund for Nature.
Boo, E. (1991a) Ecotourism: A tool for conservation and development. In J.A. Kusler (compiler) Ecotourism and Resource Conservation: A Collection of Papers (Vol. 1) (pp. 54–60). Madison: Omnipress.
Boo, E. (1991b) Planning for ecotourism. Parks 2 (3), 4–8.
Boo, E. (1992) The Ecotourism Boom: Planning for Development and Management. WHN technical paper series, Paper 2. Washington, DC: WWF.
Boo, E. (1993) Ecotourism planning for protected areas. In K. Lindberg and D.E. Hawkins (eds) Ecotourism: Guide for Planners and Managers (pp. 15–31). North Bennington: The Ecotourism Society.
Bottrill C.G. and Pearce, D.G. (1995) Ecotourism: Towards a key elements to operationalising the concept. Journal of Sustainable Tourism 3 (1), 45–54.

Amberg, N.; Magda, R. Environmental Pollution and Sustainability or the Impact of the Environmentally Conscious Measures of International Cosmetic Companies on Purchasing Organic Cosmetics. Visegrad J. Bioecon. Sustain. Dev. 2018, 1, 23.

Asadi, J., "International Environmental Labelling, Economic Consequencies, Export Magazine, July 2001

Asadi, J. 2008. Mobile Phone as management systems tools, ISO Magazine, Vol.8, No.1

Asadi, J., Eco-Labelling Standards, National Standard Magazine, Sep. 2004.

Barbieux, D.; Padula, A.D. Paths and Challenges of New Technologies: The Case of Nanotechnology-Based Cosmetics Development in Brazil. Adm. Sci. 2018, 8, 16.

Advanced Engineering and Applied Sciences: An International Journal 2014; 4(3): 26-28

Berolzheimer, C. (2006). Pencils: An Environmental Profile.

Chemical Week, 1999. Europe's Beef Ban Tests Precautionary Principle. (August 11).

Chaudri, S.K.; Jain, N.K. History of Cosmetics. Asian J. Pharm. 2009, 7–9, 164–167.

CHOI, J.P. Brand Extension as Informational Leverage. Review of Eco- nomic Studies, Vol. 65 (1998), pp. 655-669.

Conway, G. 2000. Genetically modified crops: risks and promise.

Corrado, M., (1989), The Greening Consumer in Britain, MORI, London

Corrado, M., (1997), Green Behaviour – Sustainable Trends, Sustainable Lives?, MORI, london, accessed via countries. Manila, Asian Development Bank 33p.

Davies, Clive. Chief, Design for the Environment Program, Environmental Protection Agency. Interview. March 24, 2009.

Federal Trade Commission, "Sorting Out Green Advertising Claims." http://www.ftc.gov/bcp/edu/pubs/consumer/general/gen02.shtm (March 26, 2009, March 27, 2009)

Ooyen, Carla. Research Manager with Nutrition Business Journal. Personal correspondence. March 19, 2009.

Tekin, Jenn. Marketing Manager with Packaged Facts & SBI. Personal correspondence. March 17, 2009.

University of California - Berkeley. http://berkeley.edu/news/media/releases/2006/05/22_householdchemicals.shtml (March 26, 2009)

U.S. Department of Health and Human Services, Household Products Database.http://householdproducts.nlm.nih.gov/cgi-bin/household/prodtree?prodcat=Inside+the+Home (March 17,

Women's Voices of the Earth, "Household Cleaning Products and Effects on Human Health."http://www.womenandenvironment.org/campaignsandprograms/SafeCleaning/safecleaninghealth (March 17, 2009)

EMONS, W. Credence Goods Monopolists. International Journal of In- dustrial Organization, Vol. 19 (2001), pp. 375-389.

European Union official website: https://ec.europa.eu/info/about-european-commission/contact_en

Feenstra, R.C. "Exact Hedonic Price Indexes," Review of Economics and Statistics 77 (1995): 634-653.

Feenstra, R.C., and J.A. Levinsohn. "Estimating Markups and Market Conduct with Multidimensional Product Attributes," Review of Economic Studies (62 (1995): 19-52.

ForestEthics. (n.d.). Back to School Report Card.

Forest Stewardship Council: "Principles and criteria for forest stewardship" Document 1.2: <http://www.fscoax.org>

Forsyth, K. 1999. Will consumers pay more for certified wood products? Journal of Forestry 97 (2) : 18-22.

ForestChoice #2 (2014, January 1). ForestChoice #2 Graphite Pencils (12 Pack).

Francois, C., Harris, B. (2014, November 2). How are Mechanical Pencils Made?.

Freeman, A. M III. The Measurement of Environmental and Resource Values. Theory and Methods. Washington D.C.: Resource for the Future, 1993.

Friends of the Earth, 1993. Timber certification and eco-labeling. London, FOE:

Geetha Margret Soundri, "Ecofriendly Antimicrobial Finishing of Textiles Using Natural Extract", Journal of International Academic Research For Multidisciplinary, ISSN: 2320 – 5083, 2014, Vol 2.

Graves, P., J.C. Murdoch, M.A. Thayer, and D. Waldman. "The Robustness of Hedonic Price Estimation: Urban Air Quality," Land Economics 64(1988): 220-233.

Halvorsen, R. and R. Palmquist. "The Interpretation of Dummy Variables in Semilogarithmic Equations." American Economic Review 70:474-75 (1980).

Henderson D. (2008). Opportunity Cost." The Concise Encyclopedia of Economics.

How It's Made. (2009, Nov 17). How It's Made Graphite Pencil Leads [video file].

Imhoff, Dan. "Growing Pains: Organic Cotton Tests the Fibre of Growers and Manufacturers Alike," reprinted on Simple Life's web page (simplelife.com), but first printed by Farmer to Farmer, December 1995.

Incomplete Consumer Information in Laboratory Markets. Journal of Environmental labeling.

ISO 14020, ISO 14021,ISO 14024,ISO 14025, International Organization for Standardization.

Kennedy, P.E. "Estimation with Correctly Interpreted Dummy Variables in Semilogarithmic Equations," American Economic Review 71: 801 (1981).

Kirchho®, S., (2000), Green Business and Blue Angels.

Kraus, Jeff. Lab Technician at the North Carolina School of Textiles.

Labeling Issues, Policies and Practices Worldwide.

Lamport, L. 1998. The cast of (timber) certifiers: who are they? International J. Ecoforestry 11(4): 118-122.

Large Scale impoverishment of Amazonian forests by logging and fire. 1999.

Lathrop, K.W. and Centner, T.J. 1998. Eco-labeling and ISO 14000: An analysis of US regulatory systems and issues concerning adoption of type II standards. Environmental

Lee, J. et al. 1996. Trade related environmental measures; sizing and comparing impacts.

Lehtonen, Markku. 1997. Criteria in Environmental Labeling: A comparative Analysis on Environmental Criteria in Selected Labeling Schemes. Geneva, UNEP. 148p.

LIEBI, T. Trusting Labels: A Matter of Numbers? Working Paper Uni versity of Bern, No. 0201 (2002).

OECD. "Ec-labelling: Actual Effects of Selected Programmes," OCDE/GD (97) 105, 1997, Paris. (available on line at http://www.oecd.org/env/eco/books.htm#trademono)

OECD. 1997a. Case study on eco-labeling schemes. Paris, OECD (30 Dec):

OECD. 1997b. Eco-labeling: Actual Effects of Selected Programs.

Osborne, L. "Market Structure, Hedonic Models, and the Valuation of Environmental Amenities." Unpublished Ph.D. dissertation. North Carolina State University, 1995.

Osborne, L., and V. K. Smith. "Environmental Amenities, Product Differentiation, and market Power," Mimeo, 1997.

Ozanne, L.K. and Vlosky, R.P. 1996. Wood products environmental certification: the United States perspective". Forestry Chronicle 72 (2) : 157-165.

Palmquist, R. B., F. M. Roka, and T.Vukina. "Hog Operations, Environmental Effects, and Residential Property Values," Land Economics 73(1), (1997): 114-24.

Palmquist, R.B. "Hedonic Methods," in J.B Braden and C.D. Kolstad, eds. Measuring the Demand for Environmental Improvement. Amsterdam, NL: Elsevier, 1991.

Paper Mate. (2014). Paper Mate Recycled.

Pento, T. 1997. Implementation of Public Green Procurement Programs (22-31) in Greener Purchasing: Opportunities and Innovations. Sheffield, Greenleaf Publ. 325 p.

Perloff, J. "Industrial Organization Lecture Notes," Mimeo. University of California at Berkeley (1985).

Plant, C. and Plant, J. 1991. Green business: hope or hoax? Philadelphia, New Society Publishers 136 p.

Pencil Making Today (2014, January 1). Pencil Making Today: How to Make a Pencil in 10 Steps.

Polak, J. and Bergholm, K. 1997. Eco-labeling and trade: a cooperative approach (Jan.): Policy in a Green Market. Environmental and Resource Economics 22, 419-

Poore, M.E.D. et al. 1989. No timber without trees. London, Earthscan. 352p.

Raff, D. M.G., and M. Trajtenberg. "Quality-Adjusted Prices for the American Automobile Industry: 1906-1940." NBER Working Paper Series, Working Paper No. 5035, February 1995.

Roberts, J. T. 1998. Emerging global environment standards: prospects and perils. Journal of Developing Societies 14 (1): 144-163.

Rosen, S., "Hedonic Prices and Implicit Markets: Product Differentiation in Pure Competition." Journal of Political Economy. 82: 34-55 (1974).

Ross, B. 1997. Eco-friendly procurement training course for UN HCR. : 126 p.

Sayre, D. 1996. Inside ISO 14000: The competitive advantage of environmental management. Delray Beach FL., St. Lucie Press. 232p.

Suzuki, D. (2014, January 1). PEG Compounds and their contaminants

SHAPIRO, C. Premiums for High Quality Products as Returns to Reputa- tion. Quarterly Journal of Economics, Vol. 98, No. 4 (1983), pp. 659-680.

Stillwell, M. and van Dyke, B. 1999. An activists handbook on genetically modified organisms and the WTO. Washington DC., The Consumer's Choice Council: 20 p.

Semenzato, A.; Costantini, A.; Meloni, M.; Maramaldi, G.; Meneghin, M.; Baratto, G. Formulating O/W Emulsions with Plant-Based Actives: A Stability Challenge for an Eective Product. Cosmetics 2018, 5, 59.

Sources of Plastics (2014, January 1). Sources of Plastics.

Singh, S. (2008, March 6). Paraffin wax.

Saint Jean Carbon. (n.d.). Sri Lankan Graphite.

U.S. Environmental Protection Agency. National Water Quality Fact Inventory: 1990 Report to Congress. EPA 503-9-92-006, Apr. 1992.

UK Eco-labelling Board website, accessed via http://www.ecosite.co.uk/Ecolabel-UK/

US Environmental Protection Agency (EPA742-R-99-001): 40 p. <www.epa.gov/opptintr/epp>

US EPA, 1993. Determinants of effectiveness for environmental certification and labeling programs. Washington, D.C., US Environmental Protect

US EPA, 1993. Status report on the use of environmental labels worldwide. Washington, D.C., US Environmental Protection Agency (742-R-93-001 September).

US EPA, 1993. The use of life-cycle assessment in environmental labeling. Washington, D.C., US Environmental Protection Agency (742-R-93-003 September).

US EPA, 1998. Environmental labeling: issues, policies, and practices worldwide. Washington DC., Environmental Protection Agency, Pollution Prevention Division Prepared by Abt

USG, 1998. Greening the government through waste prevention, recycling, and federal acquisition. Washington, D.C., Executive Order 13101 (September).

Kijjoa, A.; Sawangwong, P. Drugs and Cosmetics from the Sea. Mar. Drugs 2004, 2, 73–82. [CrossRef]

Wang, J.; Pan, L.; Wu, S.; Lu, L.; Xu, Y.; Zhu, Y.; Guo, M.; Zhuang, S. Recent Advances on Endocrine Disrupting Eects of UV Filters. Int. J. Environ. Res. Public Health 2016, 13, 782.

Bilal, A.I.; Tilahun, Z.; Shimels, T.; Gelan, Y.B.; Osman, E.D. Cosmetics Utilization Practice in Jigjiga Town, Eastern Ethiopia: A Community Based Cross-Sectional Study. Cosmetics 2016, 3, 40.

Ting, C.T.; Hsieh, C.M.; Chang, H.-P.; Chen, H.-S. Environmental Consciousness and Green Customer Behavior: The Moderating Roles of Incentive Mechanisms. Sustainability 2019, 11, 819.

Chen, K.; Deng, T. Research on the Green Purchase Intentions from the Perspective of Product Knowledge. Sustainability 2016, 8, 943.

Wang, H.; Ma, B.; Bai, R. How Does Green Product Knowledge Eectively Promote Green Purchase Intention? Sustainability 2019, 11, 1193.

Nguyen, T.T.H.; Yang, Z.; Nguyen, N.; Johnson, L.W.; Cao, T.K. Greenwash and Green Purchase Intention: The Mediating Role of Green Skepticism. Sustainability 2019, 11, 2653.

Cinelli, P.; Coltelli, M.B.; Signori, F.; Morganti, P.; Lazzeri, A. Cosmetic Packaging to Save the Environment: Future Perspectives. Cosmetics 2019, 6, 26.

Eixarch, H.; Wyness, L.; Siband, M. The Regulation of Personalized Cosmetics in the EU. Cosmetics 2019, 6, 29.

Appendix I: Search by Logos

Here you can search the logos in this volume. It will help you to better undersand the Ecolabels you may encounter while shopping. Buying Eco-products will aid in having a better environment with minimum polution during production processes. Three important parameteres for shopping are **quality**, **price** & **environmental impacts** of the products.

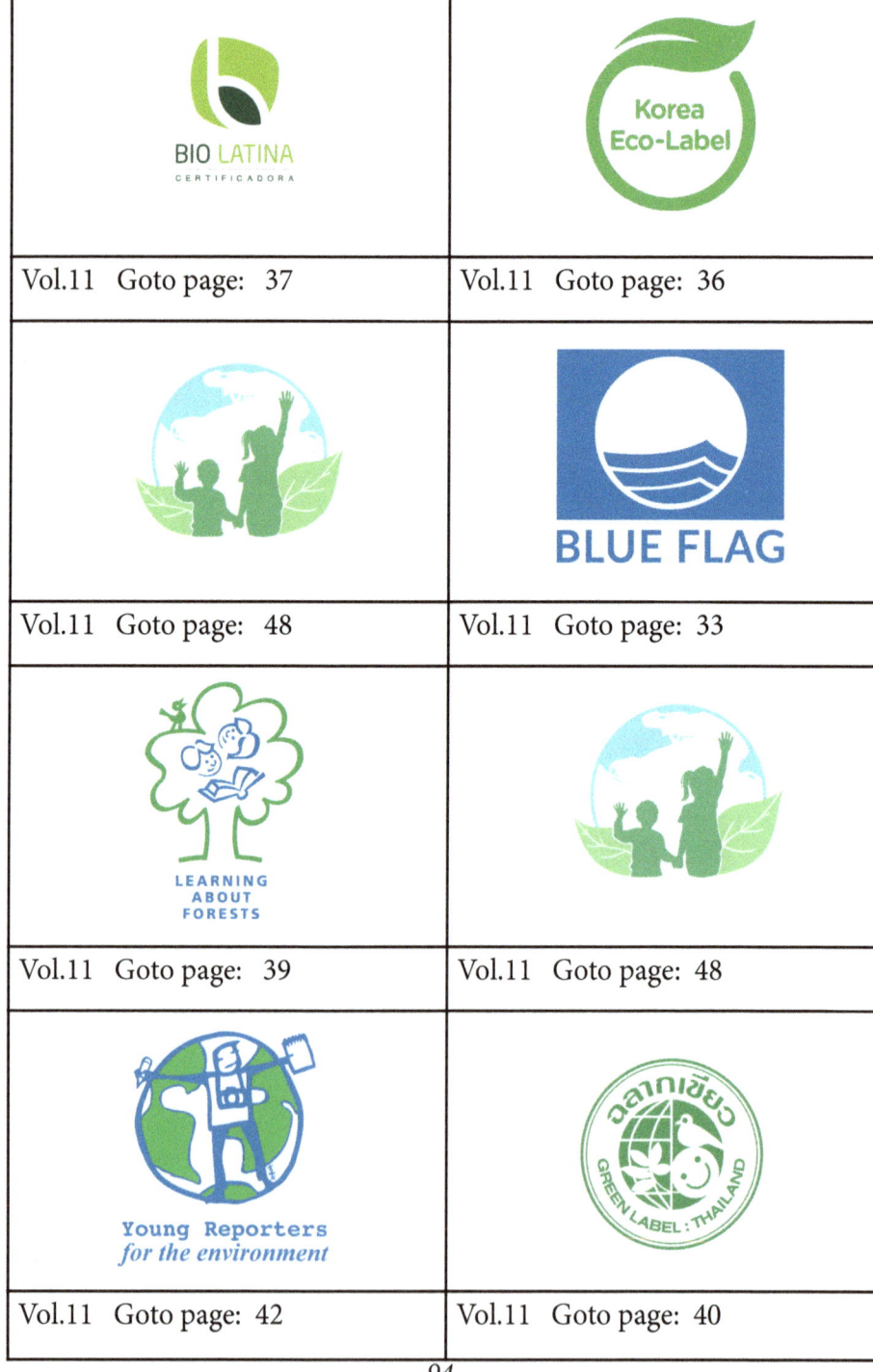

INTERNATIONAL ENVIRONMENTAL LABELLING VOL.11 • 95

Vol.11　Goto page:　30	Vol.11　Goto page:　31
Vol.11　Goto page:　42	Vol.11　Goto page:　32
Vol.11　Goto page:　39	Vol.11　Goto page:　43
Vol.11　Goto page:　48	Vol.11　Goto page:　38

Appendix II

PAPER MADE OUT OF ALGAE CELLULOSE, A SUSTAINABLE ALTERNATIVE TO CONVENTIONAL INDUSTRY

The main source of cellulose has always been wood from trees and other vascular plants. The cellulose obtained from these sources is associated with other natural polymers, mainly lignin. Algae can be considered as an alternative source of cellulose to traditional raw materials. One of the principal problems regarding the conventional extraction of cellulose is the removal of lignin. The lignin content in the algae cell wall is so low that there are not problems associated with lignin removal. Two of the most important bloom forming kinds of algae, Ulva sp. and Cladophora sp., as a raw material for papermaking. The amount of solvent-substances, lignin and holocellulose in dried algae pulp is estimated. The results show that the studied algae have low lignin-like compounds and solvent-soluble substances content, which supposes an enormous advantage over the current cellulose extraction methods as it eliminates the need of pre-treatment, cooking and bleaching stages. Therefore, the application of extremely tox-

ics reagents used nowadays is not necessary. The holocellulose content obtained , ranged from 47 to 54%, lower than that of wood or herbaceous species.
The algae genus stand as a proper source of reinforcing fibers for papermaking purposes. Also, represent an excellent opportunity to valorize tidal wastes obtained from bloom episodes.

APPENDIX III

Environmental Friendly Photos

Environmental friendly photos will be placed in this appendix. These photos can be received in the Top Ten Award International Network inbox from anywhere and everywhere, all over the globe. You can send your appropriate photos to us for them to be considered for publishing in one of the future, related volumes. They will be published with proper credit to the sender. The pictures can also be images of the Ecolabels existing in products within your country.

INTERNATIONAL ENVIRONMENTAL LABELLING VOL.11 • 99

2nd Edition

	# Vol.1 For All Food Industries (Meat, Beverage, Dairy, Bakeries, Tortilla, Grain and Oilseed, Fruit and Vegetable, Seafood, And Sugar and Confectionery)
	# Vol.2 For All Energy & Electrical Industries (Renewable Energy, Biofuels, Solar Heating & Cooling, Hydroelectric Power, Solar Power, Wind Power, Energy Conservation, Geothermal and Nuclear Power)
	# Vol.3 For All Fashion & Textile Industries (Fashion Design, The Fashion System, Fashion Retailing, Marketing and Marchandizing, Textile Design and Production, Clothing and Textile Recycling)
	# Vol.4 For All Health & Beauty Industries (Fragrances, Makeup, Cosmetics, Personal Care, Sunscreen, Toothpaste, Bathing, Nailcare & Shaving, Skin Care, Foot Care, Hair Care and Other Health & Beauty Products)

	## Vol.5 For All Maintenance & Cleaning Products (All-purpose Cleaners, Abrasive Cleaners, Powders. Liquids, Specialty Cleaners, Kitchen, Bathroom, Glass and Metal Cleaners, Bleaches, Disinfectants and Disinfectant Cleaners)
	## Vol.6 For All Wood & Stationery Industries (Wooden Products, Cardboard, Papers, Markers, Pens, NoteBooks. Writing Pads and Writing Sets, Pencils, White Papers, Envelopes and Organizers, Staplers and Paper Clips)
	## Vol.7 For All DIY & Construction Industries (Do it yourself " ("DIY") of Building, Modifying, or Repairing, Renovation, Construction Materials, Cement, Coarse Aggregates. Clay Bricks, Power Cables, Pipes and Fittings, Plywood, Tiles, Natural Flooring)
	## Vol.8 For All Agricuture & Gardening Industries (Shifting Cultivation, Nomadic Herding, Livestock Ranching, Commercial Plantations, Mixed Farming, Horticulture, Butterfly Gardens, Container Gardening, Demonstration Gardens, Organic Gardening)

2nd Edition

	## Vol.9 For All Professional Products & Services (Teachers, Pilots, Lawyers, Advertising Professionals, Architects, Accountants, Engineers, Consultants, Human Resources Specialist, R&D, Psychologists, Pharmacist, Commercial Banker, Research Analyst)
	## Vol.10 For All Financial Products & Services (Banking, Professional Advisory, Wealth Management, Mutual Funds, Insurance, Stock Market, Treasury/Debt Instruments, Tax/Audit Consulting, Capital Restructuring, Portfolio Management)
	## Vol.11 For All Tourism Industries (Airline Industry, Travel Agent, Car Rental, Water Transport, Coach Services, Railway, Spacecraft, Hotels, Shared Accommodation, Camping, Bed & Breakfast, Cruises, Tour Operators)
	**Set Box Books Vol.1-11** **+ Free Knowledge Test** for Schools, Libraries, Homes and Offices all over the globe: www.TopTenAward.Net

www.ingramcontent.com/pod-product-compliance
Lightning Source LLC
Chambersburg PA
CBHW040421100526
44589CB00021B/2780